Priority Areas for Research in Homeopathy

DR. SAURAV ARORA

DR. BHARTI ARORA

ISBN-13: 978-1507723753

ISBN-10: 150772375X

DEDICATION

Dedicated to all practitioners, researchers, promoters, and lovers of Homeopathy.

CONTENTS

ACKNOWLEDGMENT

The sincere efforts, contributions, and suggestions of the following members, who have extended their valuable time and expertise are duly acknowledged.

In alphabetical order of their first name:

Anisur Rahman Khuda-Bukhsh, Ph.D., Retired Professor, Department of Zoology, University of Kalyani, India.

D. R. Lohar, M.Sc., Ph.D. Organic Chemistry, Former Director, Homoeopathic Pharmacopoeia Laboratory, India.

Emilio Cervera, Associate Professor (Medicine), Complutense University of Madrid and Francisco de Vitoria of Pozuelo, Madrid.

George Vithoulkas, International Academy of Classical Homeopathy, Greece.

Hugh Harrison, Watchbell House Natural Therapy Centre, United Kingdom.

J. P. Varshney, Retd. Professor Veterinary Medicine, IVRI Izatnagar, Senior Consultant, Nandini Veterinary Hospital, Surat, India.

Joyce Frye, Science Editor, American Journal of Homeopathic Medicine.

Jürgen Clausen, Karl und Veronica Carstens-Stiftung, Germany.

Kinjal Shah, MD (Hom.), Mumbai, India.

Leoni V. Bonamin, Vet MD, Ph.D., Fellow of CNPq Research Productivity – Level 2, Brazil.

Lex Rutten, MD, Independent Researcher, Netherlands.

Michel Van Wassenhoven, President, Belgian Homeopathic Medicines Registration Commission AFMPS, Belgium.

Nick Thompson, BSc (Hons) Path Sci., BVM&S, Vet MF Hom, MRCVS, Holisticvet Ltd, United Kingdom.

Omar Matuk, Assistant Medical Director, Legacy Community Health Services Voluntary Faculty, Texas Children Hospital, Houston.

Priya Kamath, BHMS, voluntary worker, Great Western Hospital, Swindon, England.

R. Valavan, Manager Scientific Affairs, Willmar Schwabe India Pvt. Ltd., India.

Sandeep Saluja, Post Graduate (Internal Medicine), Volunteer worker, India.

Seyedaghanoor Sadeghi, Medical Doctor and Homeopath, Fahimeh Mahdevi.

Shahram Shahabi, MD, Ph.D., Associate Professor, Department of Immunology, Faculty of Medicine, Urmia University of Medical

Sciences, Iran.

Shantaram G. Kane, Adjunct Professor, Indian Institute of Technology, Bombay, India.

Shivang Swaminarayan, Head-healthcare Division, Sintex International Ltd., India.

Silvia Waisse, Professor, Post Graduate Program in History of Science, PUC-SP, Brazil; Researcher, Center Simao Mathias for the History of Science (CESIMA), PUC-SP, Brazil; Executive Editor, International Journal of High Dilution Research.

Swapna Potdar, MD (Hom.), Pune, India.

Uta Mittelstadt, MSc (Homeopathic medicine), University of Central Lancashire.

1 WHY RESEARCH IN HOMEOPATHY

Homeopathy has gained a universal popularity because of its effectiveness, affordability, availability, easier applicability and no side effects. It is now the second largest practiced system of medicine worldwide. On the other hand, it has always been questioned for use of ultra-high dilutions, potentization, and biological actions, and also for its philosophy, prescription modalities, and standardization issues. Therefore, it has been dragged amongst the most controversial issues of this century.

Since inception, it is compared with the conventional system of medicine because the modern pharmacology "knows" physical and chemical interactions between drugs and living cells, whereas it is assumed that homeopathy has no answer in this area. For many decades, it was difficult to define the exact 'paths' from medicine to biology, then to chemistry and physics, but conventional systems have taken an edge over it, whereas it is considered that homeopathy still has to prove these paths.

The controversy and priority issues have always ignited intellectual minds to explore "whether homeopathy works and if yes, how does it work?", "what is nature ultra-high dilutions" and many similar questions. Till date a lot of research has been carried out to explore ultra-high dilutions from various aspects be it homeopathy; fundamental, clinical, agronomy, veterinary or dentistry research; or their usage in conventional research.

Despite advances, research in homeopathy is still a challenge, attributed to a limited number of research studies (with respect to conventional research), complex nature of ultra-high dilutions, inadequate exposure, limited resources, inaccessibility to published literature, poor motivation, lack of experience, jealousy etc.

Therefore, research is a constant need to explore, establish, answer and support homeopathy!

2 NEED FOR DOCUMENT ON PRIORITY AREAS FOR RESEARCH IN HOMEOPATHY

The surge in research has sparked a new era, where the responsibility lies on undertaking quality research. There is a small but robust research portfolio for homeopathy ranging from basic science to clinical research and it is continuing to grow in other fields also. To be able to do research in a methodical way one needs a good research question. Although a research question may be generated out of choice, interest, observation, and uncertainty but still there are many areas where we should focus our resources in order to attain quality documentation of reliable evidence. Priority areas also serve as a guide for students, researchers and funding agencies. The aim of prioritization is to promote innovative proposals to stimulate increased quality, visibility or availability of research at a national, regional or international level. The "*priority*" also depends on many factors which may range from available resources, the applicability of research, health care needs, expertise, etc., some of these are considered in development of this document.

3 ABOUT IPRH

Foundation of Initiative to Promote Research in Homeopathy (IPRH)

Research in homeopathy has always been a challenge for everyone, be it a researcher or a layman. Various reasons like complex nature of ultra-high dilutions, homeopathic philosophy, lack of apt research orientation, inadequate research literacy, lack of funds, poor motivation, skepticism etc., have always discouraged many to undertake research. The magnitude of existing research is considered insufficient to cater the need of the hour. Also, various research works have been done, but either they are not understood enough or less quoted or both.

It is also true that we may overcome most of the hindrances by sharing knowledge, expertise, and resources.

The Initiative to Promote Research in Homeopathy (IPRH) was founded, formalized and presented to the profession in the year 2011 by Dr. Saurav Arora. The aim was to enhance research awareness, disseminate scientific research, knowledge exchange and extending help for undertaking research in homeopathy through IPRH.

"IPRH runs on the philosophy of liberal exchange of scientific knowledge and expertise."

IPRH key objectives

- Promotion of research education and sensitizing people about research literacy, understanding, and expertise.
- Exchange of knowledge, expertise and resources.
- Capacity building by organizing CMEs/seminars/workshops for persons desirous of undertaking research in homeopathy.
- Formulation and maintenance of knowledge databases e.g. Homeopathic Research Database.
- Identification of priority areas of research in Homeopathy.
- Collaboration with institutions, organizations and individuals for undertaking research in homeopathy.
- Formulation of generic protocols for clinicians.

- Development of guidelines/resources/handy information about research.
- Extending help for undertaking research in homeopathy.

This initiative is beyond the boundaries of an institution and has created a change in research perspectives.

Projects by IPRH

- **Homeopathic Research Database** compiles peer-reviewed scientific publication in homeopathy and high dilution.
- Promotion of Homeopathic Research by Community Initiated and Supported Studies (**PHORECISS**)
- **Research Updates – Homeopathy**: An online, quarterly e-journal dedicated to homeopathy.
- **Research Material**: It is a constant effort of IPRH to develop quality and authentic research material for researchers and clinicians.
- **Research Literacy**: IPRH firmly believes that research literacy is the first step towards research.
- **Surveys**

More information about IPRH is available at www.researchinhomeopathy.org

4 APPLICABILITY OF THIS DOCUMENT

Priority Areas for Research in Homeopathy (PARH) doesn't aim to replace the existing documents/material on priority areas developed by institutions, researchers and health care systems, but rather it is aimed to aid them with a common consensus of global experts. This document may be used to:

- Create awareness amongst homeopathic fraternity regarding research areas which are of vital importance considering present status of homeopathy, resources, and expertise.
- Ease out identification of areas of one's own interest and thus provoke thinking process regarding "where to start working".
- Suggest an exhaustive list of priority areas which are important to venture into, and thus innovative proposals may be developed on these lines.

5 EXECUTIVE SUMMARY OF PRIORITY AREAS FOR RESEARCH IN HOMEOPATHY DOCUMENT

In May 2014, IPRH published the first version (1.1) of **Priority Areas for Research in Homeopathy** (PARH) whose development was guided by common questions people usually ask, *"where and what to start researching in homeopathy"*.

This document received appreciation as well as criticism. In November 2014 the second version (1.2) was published considering the reviews received for version 1.1.

The current version (1.3) preserves the original frameworks (1.1 and 1.2) with minor updates, as deemed necessary.

6 **OBJECTIVES**

The PARH has been developed to:-

"Identify homeopathic research priorities considering experiences and critical gaps that can be adapted by clinicians, academicians, researchers/research institutions, local health authorities and funding agencies".

7 **METHODS**

The methodology included:

1. *Collation of existing key documents.*
2. *Identification of participants comprised of teachers, researchers, clinicians, policy makers, etc.*
3. *Determination of the research priorities:*
   a. *Round One* – participants were asked to provide their suggestions and inputs to a draft document prepared beforehand (what they believed to be a priority).
   b. *Round Two* – the document generated from the round one was sent to participants who were then asked to accept or reject inputs.
4. *Validation of the draft document* – the document was then validated for inputs to which all members agreed.

The use of the electronic media (emails and the IPRH website) was a central feature in the development of the document.

8 THE NEXT STEP

This work is an ongoing process, where readers are invited to comment, criticize and contribute. If you like to contribute any addition or deletions you may get in touch with Dr. Saurav Arora at info@researchinhomeopathy.org

9 PRIORITY AREAS

Homeopathic Principles

⊕ Understanding of modus operandi of homeopathy[1, 2] (mechanism of action of homeopathic medicine) in biological/physical models keeping proper controls.

⊕ Demonstration of the law of similar[1] and to elicit its recognizable biological effects.

⊕ To understand the action of succussion and trituration on the vehicle, a vehicle containing drug and how it differs from only diluted vehicle/vehicle containing drug (unsuccussed)[1].

⊕ Independent reproduction of existing, high-quality fundamental research experiments.

⊕ Exploration and demonstration of miasmatic theory, Hering's Law of Cure, Kent's 12 observations by epidemiological tools and scientific models and their relevance and applicability to clinical practices.

⊕ Understanding the concept related to "right" dosage, frequency and optimization of repetition of homeopathic medicines in different potencies.

⊕ Homeopathy as personalized medicine keeping in view "prognostic model" factor.

⊕ To understand homeopathy as a bio-semiotic system: e.g. the role of Body Signifier Theory from M Bastide and A Lagache and other related theories.

⊕ Surveys regarding
  o Usage of potencies in various scales e.g. Decimal, centesimal, LM etc.
  o Usage of nosodes, sarcodes, imponderabilia and tautopathy.

Fundamental Research

℘ An exploration into physicochemical nature of homeopathic medicines, and understand how they differ from their vehicle[1, 2] and how much such differences are related to the corresponding biological effects.

℘ Modulation of biological pathways by homeopathic medicines (supra- cellular and sub-cellular mechanisms etc.).

℘ Understanding of the systemic features of homeopathy effects in vivo and modulation of the neural-immune-endocrine net.

℘ Influence of various factors on homeopathic medicines[1] e.g.
  o Light
  o Temperature
  o Vibration
  o Electromagnetic field
  o Transportation
  o Storage container[1]
  o Ageing
  o Food items such as coffee/tea/strong smelling objects etc.

℘ The presence of aqueous nanostructures, their nature and biological action and relevance to the homeopathic medicines effects.

℘ Nano-encapsulation of mother tinctures of homeopathic drugs and specific nano-precipitation ability of homeopathic drugs.

℘ Genomic/Epigenetic[1]/phenotypic studies in in-vitro and in-vivo models.

℘ Studies for the standardization of preparation of homeopathic medicines (trituration and succussion processes), and effect of different procedure of succussion (e.g. sonication, hand jerks, vortexing) and containers (like glass, polypropylene etc.) and to optimize the procedure that could give best results.

℘ To understand the role of alcohol and water in homeopathic potencies, as a solvent.

℘ Transfer of biological information from liquid to globules and to living organism.

Clinical research

Prevalence of diseases and symptoms in:

- Populations responding well to specific homeopathic medicines.
- The whole population.

Disease conditions - categorized primarily as per ICD classification

℞ *Cardiovascular disorders*
  o Moderate to severe functional disorders.
  o Absorption of intravascular clots in patients post heart attack.
℞ *Disease of circulatory system*
  o Cerebrovascular disorders

- ℞   o Apoplexy
- ℞ *Endocrine, Nutritional & Metabolic disorders*
  - o Diabetes mellitus[1]
  - o Subclinical thyroid disorders[2] especially subclinical hypothyroidism and autoimmune thyroiditis
  - o Wilson's
  - o Polycystic ovarian syndrome
- ℞ *Diseases of digestive disorders*
  - o Acid peptic disorders[2]
  - o Irritable bowel syndrome[2]
  - o Ulcerative Colitis[2]
  - o Crohn's disease
- ℞ *Diseases of Skin*
  - o Trophic ulcer[2]
  - o Vitiligo[2]
  - o Psoriasis
- ℞ *Genitourinary disorders*
  - o Dysmenorrhoea[2]
  - o Premenstrual syndrome[2]
  - o Benign hyperplasia of prostate[2]
  - o Infertility
  - o Exploration of definite contraceptive effects, if any.
- ℞ *Disorders of Musculoskeletal system*
  - o Osteopenia/Osteoarthritis
  - o Arthralgia
  - o Arthritis
- ℞ *Disease of nervous system*
  - o Alzheimer's[1]
  - o Motor Neuron Disease[1]
  - o Parkinsonism[1]
- ℞ *Mental and Behavioural disorders*
  - o Behavioural disorders in pediatric population
  - o Depression[2] and anxiety
  - o Autism[2]
  - o Attention deficit hyperactive disorders
  - o Dementia[1]
- ℞ *Diseases of Respiratory system*
  - o The recurrent tendency to catch upper respiratory tract infections.
  - o Influenza-like illnesses[2]
  - o Asthma (Allergic and Bronchial)[2]
- ℞ *Neoplasms*

- o Homeopathy as a tool for improving quality of life in oncologic patients receiving chemotherapy or post chemotherapy
- ℞ *Infectious and parasitic diseases*
  - o Herpes zoster[2]
  - o Japanese Encephalitis[1]
  - o Dengue[1,2]
  - o Malaria[1]
  - o Chikungunya[1,2]
  - o HIV/AIDS[1,2]
  - o MDR Tuberculosis[1]
  - o Leishmaniasis
- ℞ *Mother and Child care*
  - o Safety of homeopathic medicines in pregnancy and new born/infants

Using paramedical tools

- ☐ Using laboratory tests and imaging tools (e.g. plain X-rays, sonography, CT scan, MRI, etc.) to distinguish proving-induced changes.
- ☐ Using laboratory tests and imaging tools (e.g. plain X-rays, sonography, CT scan, MRI, etc.) to help empirical choosing of suitable remedies for acute conditions and special symptoms.

Drug development and standardization

- ❖ Identification of shelf life of homeopathic medicines
  - *a. Mother tinctures*
    1. Shelf-life of mother tincture and mother solution
    2. In different plastic containers vs glass containers
  - *b. Dilutions and medicated pills*
    1. In different plastic containers vs glass containers
- ❖ Development of guidelines, checklists and implementation strategies for standardization and quality assurance[1].
- ❖ Research related to techniques used in pharmaceutical industry and drug development[1].
- ❖ Technological issues regarding the preparation of homeopathic medicines[1].
- ❖ Pharmacological studies[1] (Safety and efficacy).
- ❖ Efficacy of homeopathic medicines manufactured using different potentisers.

❖ Efficacy of homeopathic medicines manufactured using different techniques like Hahnemannian and Korsakovian methods.

Drug Proving

∂ Changes in the laboratory parameters in addition to symptoms obtained during drug proving.
∂ Identification of techniques to use laboratory parameters used during proving to clinical application/prescriptions.

Homeopathy in Public Health

Homeopathy is practiced in many countries as a part of public health system. The health authorities of each country/region have the power to identify priority areas for research in homeopathy. The prioritization depends on many factors, some of which may be resources, prevalent diseases and unanswered questions to medical problems which otherwise cannot be dealt by mainstream medical practices. There is a need to identify global as well as local priority areas for research in homeopathy, especially for public health use. This may be notified to health authorities for efficient implementation of policies and undertaking research in areas from time to time. Following are some of the areas which may be considered for public health research:

1. Integrated research in communicable diseases, vector control studies, control of reservoirs of disease and surveillance of communicable diseases.
2. Amelioration of natural toxicity where orthodox systems of medicine fail: e.g. arsenic toxicity, fluoride toxicity, mercury toxicity, etc.
3. Delaying toxicity of snake venom or prophylactic application for prevention/delay in toxic effects of snake and another animal venom (insects, arthropods etc.)
4. Anaemia.
5. School absenteeism.
6. Knowledge, Attitude and Practices (KAP) in preventive/community medicine.
   a. The effectiveness of homeopathic medicines for prevention of common diseases.
   b. Genus epidemicus for commonly/frequently occurring endemic/pandemic diseases.
7. Research related to immunization of diseases which are not notified as essential immunizations in a public health system (optional or short term vaccinations).

8. Surveys related to the use of homeopathy in public health system, its applicability, acceptability, usage etc.

Agro-Homeopathy

- Studies related to enhancing plant growth and productivity.
- Identification of indications of medicines to be used in plant diseases.
- Treatment strategies for using homeopathy in plant diseases.
- The inclusion of homeopathic preparations in pharmacopoeia for their use in agricultural industry.
- Development of plant models for basic research.
- Efficacy of judicious application of homeopathic drugs in large-scale forestry.

Veterinary Homeopathy

- Disease related to decreased productivity in animals[1].
- To enhance productivity and quality of life of farm animals[1] and fishes.
- Development of nosode(s) for preventing infectious diseases as well as control of ticks and mites.
- Mastitis.
- Diarrhoea[1].
- Skin diseases[1].
- Reproductive disorders (metritis, infertility etc.)[1].
- Wound/Injury[1]/Infection management.
- Infectious diseases, specially zoonosis and understanding the particularities of host-parasite mechanisms under homeopathic treatment.
- Homeopathy in senescent animals.
- Homeopathy as a tool of animal welfare.
- Laminitis and joint diseases in horses.
- Pain management in animals.
- Osteoarthritis and hip dysplasia management in dogs.
- Urolith management in animals.

Homeopathic market research

- ✓ Market size of homeopathic industry
  - o Pharmaceutical industry.

- o Corporate clinics.
- o Private clinics.
- o Government sector.
- o Market size based on countries/regions/etc.
- o Domestic, import and export business.
- ✓ Market share of different pharmaceutical companies
  - o Ratio of imported vs domestic products sale
- ✓ Market share of top generic brands like
  - o Variants of general health tonics of different companies
  - o Weight reducing products
  - o Heart tonics
  - o Bio-chemics and Bio-combinations
  - o Hair oils
  - o Etc.
- ✓ The ratio of Over the Counter (OTC) Sale vs prescription vs sale to the physician.
- ✓ The ratio of single remedies vs branded products.
- ✓ Study on the ratio of expenditure in homeopathic dispensaries/clinics under various heads viz. dilutions, mother tinctures, globules, lactose, sundries, branded products, etc.
- ✓ Explorative study on type of practitioner
  - o Classical
    - ▪ Single remedy and single dose
    - ▪ Single remedy and multiple doses
  - o Non-classical
    - ▪ Single remedy + Bio-chemics
    - ▪ Single remedy + Bio-chemic + mother tinctures
    - ▪ Multiple remedies + Bio-chemics
    - ▪ Multiple remedies + Bio-chemic + Bio-combination + mother tinctures
    - ▪ Multiple remedies + biochemic + biocombination + mother tinctures + branded products

✓ Use of homeopathic medicine as home remedy/self-medication.

10 **REFERENCES**

1. Areas, Guidelines/Operative procedures of Expression of Interest for Fundamental & Collaborative Research [Internet] 2014 [cited 2014 October 24]; pp 4-5. Available from: ccrhindia.org/pdf/EoI-CCRH-Scope_Format_Application.pdf
2. Extra Mural Research Scheme of Department of AYUSH, Govt. of India. [Internet] 2014 [cited 2014 October 24]; pp 21-22. Available from http://indianmedicine.nic.in/writereaddata/linkimages/1682595779-Revised%20Scheme%20of%20EMR%2024.pdf
3. An interview with Dr. Van Wassenhoven. Available online at http://www.echamp.eu/news/newsletter/newsletter-archive/2011/september/an-interview-with-dr-van-wassenhoven.html. Last accessed on March 14, 2014.

For Disclaimer see www.researchinhomeoapthy.org

ABOUT THE AUTHORS

Dr. Saurav Arora, Gold Medalist, is an internationally acclaimed homeopathic physician, consultant, and researcher. Dr. Arora is currently the Editor-in-chief of International Journal of High Dilution Research, Director of Initiative to Promote Research in Homeopathy and Director of the Registration Society of Homoeopathy, UK. To know more about Dr. Saurav Arora you may visit www.sauravarora.com

Dr. Bharti Arora, Silver Medalist, is a practitioner and promoter of Homeopathy, committed to promoting it for the benefit of individual and health care system. Dr. Bharti is the co-founder of Vivid Homeopathy and Arora's Homeopathic Clinic and provides consultancy at two clinics in New Delhi. To know more about Dr. Bharti Arora you may visit www.arorasclinic.com

www.ingramcontent.com/pod-product-compliance
Lightning Source LLC
Chambersburg PA
CBHW041617180526
45159CB00002BC/903